3 Nebraska NSCAS Grade 3 Math Practice Tests

Full-Length Test Prep with Detailed Answer Explanations

Dr. A. Nazari

3 Practice Tests to Get You Started!

Hey there, future math whiz!

This book has **3 full practice tests** to help you warm up for the real thing. Think of it like stretching before a big game — these tests will get your brain ready and show you what to expect!

👍 Three tests is the **perfect start**!

👍 Each one helps you feel **more ready**!

👍 You'll be surprised how much you **already know**!

Sharpen your pencil and let's get warmed up! 🔥

> **❝** Three practice tests is a great way to start. Take your time with each one, and you'll feel more confident every step of the way! **❞**

 How to Use This Book

- *3 Full-Length Practice Tests* — Each one covers all the Grade 3 math topics you need to know!

- *Answer Key with Explanations* — Find out why each answer is correct, not just what the answer is.

- *Reference Pages* — A math symbols chart and multiplication table you can peek at any time.

- *A Test Tracker* — Write down your scores and watch your confidence grow!

With just 3 tests, here's a great way to use them:

- *Test 1*   **The Warm-Up.** Take this test without a timer. Get comfortable with the question types. Don't worry about your score — just do your best!

- *Test 2*   **The Practice Round.** After reviewing Test 1, try this one with a timer (ask a grown-up!). Focus on the topics that were tricky last time.

- *Test 3*   **The Real Deal.** Treat this like the actual test: quiet room, timed, no peeking at answers. See how much you've improved!

Pick the **one best answer** from choices A, B, C, or D. Not sure? Cross out the ones you know are wrong, then pick from what's left. That's a smart move!

Write your answer **and** show your work! Even if your final answer isn't right, showing your steps can earn you credit. Use scratch paper if you need more room.

❝ After Each Test ❞

Flip to the Answer Key and check your work. For every question you got wrong, **read the explanation carefully**. Then write the tricky topics on your Test Tracker page. If you need extra help, grab our *Grade 3 Math Study Guide!*

⭐ ***Fun fact:*** *Three tests is all it takes to see real improvement! Most kids feel way more confident after just a few rounds of practice.* ⭐

Find more at
ViewMath.com/NE-Grade3

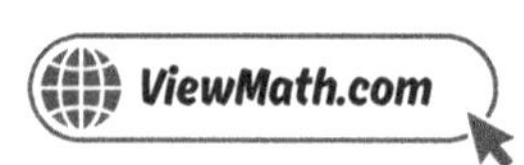

💡 Tips for Test Day 💡

- ✅ **Sleep early** — *your brain learns while you sleep!*
- ✅ **Pack your supplies** — *pencils, eraser, scratch paper, all ready to go.*
- ✅ **Tell yourself:** *"I've been practicing. I'm going to do great!"*

5 Simple Rules for Every Test

1. **Read the question twice.** *The first time to understand it. The second time to catch details.*
2. **Show your work.** *Write the steps down, even on scratch paper. It helps you think!*
3. **Skip the hard ones.** *Put a small star next to tricky questions and come back later. Answer the easy ones first!*
4. **Never leave a blank.** *For multiple choice, your best guess is better than no answer at all.*
5. **Check your work.** *Finished early? Go back and re-read your answers.*

- Take a deep breath before you begin
- Underline key words in the question
- Use drawings or number lines to help
- Cross out wrong answers first
- Double-check addition and subtraction

- Rushing and not reading carefully
- Picking the first answer that "looks right"
- Forgetting to carry or borrow numbers
- Skipping a question permanently
- Panicking when you see a tough problem

" Remember, the very first practice test is the hardest — not because the questions are harder, but because everything is new! By Test 3, you'll feel like a pro. Trust me! "

# Get Ready to Practice

Here's everything you need before you start!

Pencils

Sharpened and ready!

Eraser

Everyone makes mistakes!

Scratch Paper

For working things out

A Calm Spot

Somewhere quiet to focus

A Grown-Up

To help set a timer

A Can-Do Attitude

You've totally got this!

Allowed During Tests

- Pencils and erasers
- Blank scratch paper
- The **reference pages** in this book
- A ruler (for measurement questions)

Not Allowed

- Calculators
- Phones, tablets, or computers
- Help from anyone else
- Your study guide (save it for after!)

For Parents & Teachers

- With only 3 tests, **space them at least a week apart**. This gives time to review mistakes before trying the next one.
- Let your child take Test 1 untimed to build familiarity.
- After each test, go through the Answer Key together. Focus on **understanding the "why,"** not just the score.
- If a topic keeps tripping them up, review it in our **Grade 3 Math Study Guide** before the next practice test.
- Celebrate every bit of progress — even getting one more question right is a win!

X^1 Math Reference Sheet X^1

Symbol	Name	What It Means	
$+$	Plus (Add)	Put numbers together.	$3 + 5 = 8$
$-$	Minus (Subtract)	Take away from a number.	$9 - 4 = 5$
$\times$	Times (Multiply)	Add equal groups.	$4 \times 3 = 12$
$\div$	Divide	Split into equal groups.	$12 \div 3 = 4$
$=$	Equals	Both sides are the same.	$2 + 3 = 5$
$>$	Greater Than	The left number is bigger.	$7 > 3$
$<$	Less Than	The left number is smaller.	$2 < 9$
$\frac{}{}$	Fraction Bar	Part of a whole.	$\frac{1}{2}$ means 1 out of 2 equal parts

- **Sum** — the answer when you add
- **Difference** — the answer when you subtract
- **Product** — the answer when you multiply
- **Quotient** — the answer when you divide
- **Factor** — a number you multiply
- **Array** — objects in rows and columns
- **Fraction** — a part of a whole

- **Numerator** — the top number in a fraction
- **Denominator** — the bottom number
- **Equation** — a math sentence with $=$
- **Estimate** — a smart guess, close to the real answer
- **Perimeter** — the distance around a shape
- **Area** — the space inside a shape
- **Rounding** — making a number simpler by going to the nearest ten or hundred

- **Add** (+): in all, total, altogether, combined, sum, both, more
- **Subtract** (−): how many more, how many left, fewer, difference, remain
- **Multiply** (×): each, every, groups of, times, rows of, per
- **Divide** (÷): share equally, split, each group, how many groups, per

Multiplication Table

×	1	2	3	4	5	6	7	8	9	10	11
1	1	2	3	4	5	6	7	8	9	10	11
2	2	4	6	8	10	12	14	16	18	20	22
3	3	6	9	12	15	18	21	24	27	30	33
4	4	8	12	16	20	24	28	32	36	40	44
5	5	10	15	20	25	30	35	40	45	50	55
6	6	12	18	24	30	36	42	48	54	60	66
7	7	14	21	28	35	42	49	56	63	70	77
8	8	16	24	32	40	48	56	64	72	80	88
9	9	18	27	36	45	54	63	72	81	90	99
10	10	20	30	40	50	60	70	80	90	100	110
11	11	22	33	44	55	66	77	88	99	110	121

To find **4 × 7**:

1. Find **4** in the left column (blue).
2. Find **7** in the top row (blue).
3. Follow the row and column until they meet: the answer is **28**!

 My Confidence Tracker

Record your scores below. You'll be amazed at your progress!

My name: ________________________________

☑ Test	📅 Date	⭐ Score	😊 How I Feel
1			
2			
3			

The easiest topic for me was:

____

The trickiest topic for me was:

____

One thing I got better at from Test 1 to Test 3:

____

Next time I want to try:

____

You just finished 3 practice tests — that's awesome! Compare your first score to your last. I bet you'll see real improvement. Ready for more? Check out our 5-test or 7-test books for even more practice!

Find more at
ViewMath.com/NE-Grade3

Table of Contents

Let's learn and have fun!

Practice Test 1

30 Questions

✏ Before You Start ✏

- ✓ **Read each question carefully** before choosing your answer.
- ✓ **Show your work** on scratch paper when you need to.
- ✓ **Skip hard questions** and come back to them later.
- ✓ **Check your answers** when you're done.
- ✓ **Take your time** — there's no rush!

⭐ You've Got This! ⭐

Do your best and show what you know!

Write 5,600 *in expanded form.*

Which number rounds to 500 when rounded to the nearest 100?

A 428

B 449

C 462

D 551

3. Which group contains only even numbers?

A 12, 34, 57

B 20, 46, 88

C 31, 50, 72

D 14, 63, 90

A school raised $8,000 *last year and* $2,000 *this year. How much did the school raise in total?*

A $6,000

B $10,000

C $82,000

D $28,000

What is the value of the digit 8 in 283,041?

A 80

B 800

C 8,000

D 80,000

6. What is $6,789 + 3,456$?

7. *A farmer harvested 7,204 apples. He sold 3,568 apples at the market. How many apples does the farmer have left?*

8. *When you estimate $347 + 582$ by rounding each number to the nearest hundred, the estimate is 900. The exact sum is 929. Which statement is true?*

- A *The estimate is greater than the exact sum*
- B *The estimate is less than the exact sum*
- C *The estimate equals the exact sum*
- D *You cannot compare an estimate to an exact answer*

9. *I am a number. When you divide me by 8, you get 9. What number am I?*

10. *Which set of numbers does NOT form a fact family?*

- A $3, 9, 27$
- B $6, 7, 42$
- C $4, 5, 25$
- D $8, 8, 64$

11. *In the fraction $\frac{5}{6}$, what does the 5 tell you?*

- A *There are 5 equal parts in the whole*
- B *You need 5 more parts*
- C *You have 5 of the equal parts*
- D *The whole has 5 pieces left*

Get Online

Find more at
ViewMath.com/NE-Grade3

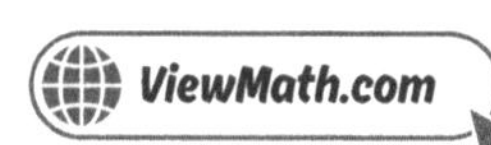

12. A number line is divided into 4 equal parts. Which fraction is at the halfway point between 0 and 1?

A $\frac{1}{4}$

B $\frac{2}{4}$

C $\frac{3}{4}$

D $\frac{4}{4}$

5 friends share a sandwich equally. What unit fraction does each friend get?

How many copies of $\frac{1}{3}$ make 1 whole?

A 1

B 2

C 3

D 4

15. $\frac{2}{4} = \frac{?}{2}$. What is the missing numerator?

Write 4 as a fraction with denominator 3.

A $\frac{4}{3}$

B $\frac{3}{4}$

C $\frac{7}{3}$

D $\frac{12}{3}$

Tom ate $\frac{3}{8}$ of a pizza. Mia ate $\frac{5}{8}$ of the same pizza. Who ate more?

A Tom

B Mia

C They ate the same amount

D Cannot tell

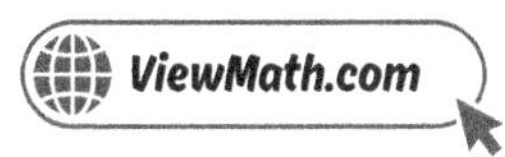

18. What is $\frac{3}{4} + \frac{1}{4}$?

19. When you subtract fractions with the same denominator, what do you do?

   A Subtract both the numerators and the denominators.

   B Subtract the numerators and keep the denominator the same.

   C Subtract the denominators and keep the numerator the same.

   D Multiply the numerators.

20. A string is $6\frac{1}{4}$ inches long. A ribbon is $6\frac{3}{4}$ inches long. How much longer is the ribbon than the string?

   A $\frac{1}{4}$ inch

   B $\frac{1}{2}$ inch

   C $\frac{3}{4}$ inch

   D 1 inch

21. A bag of apples has a mass of 3 kg. A bag of oranges has a mass of 5 kg. What is the total mass?

   A 2 kg

   B 15 kg

   C 8 kg

   D 35 kg

22. A large milk jug holds 4 liters. You pour out 1 liter for cereal. How much milk is left?

23. What is $\$2.50 + \$1.25$?

   A $3.25

   B $3.75

   C $4.75

   D $3.55

Find more at
ViewMath.com/NE-Grade3

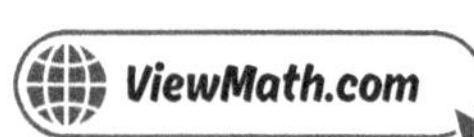

24. *A picture graph has the key: Each* ★ *= 5 books. The row for Jake has 3 stars. How many books did Jake read?*

A 3

B 5

C 8

D 15

25. *A line plot shows pencil lengths. There are 3 X marks above $1\frac{1}{2}$ inches. What does this mean?*

A 3 pencils are $1\frac{1}{2}$ inches long

B $1\frac{1}{2}$ pencils are 3 inches long

C The total length is $4\frac{1}{2}$ inches

D There are 3 pencils in all

26. *A bag has ONLY red marbles. What is the probability of picking a blue marble?*

A Certain

B Likely

C Unlikely

D Impossible

27. *A regular polygon has all sides the same length and all angles the same size. Which of these is always a regular polygon?*

A Rectangle

B Trapezoid

C Square

D Rhombus

28. *A square has sides that are 7 inches long. What is its area?*

A 14 sq in

B 28 sq in

C 42 sq in

D 49 sq in

Find more at
ViewMath.com/NE-Grade3

29. *A pizza is cut into 2 equal slices. What fraction is each slice?*

    (A) $\frac{1}{2}$ (B) $\frac{1}{3}$

    (C) $\frac{1}{4}$ (D) $\frac{2}{2}$

30. *A STOP sign is shaped like a regular.*

    (A) Hexagon   (B) Heptagon

    (C) Octagon   (D) Decagon

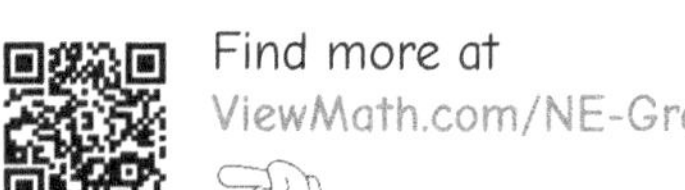

# ⭐ *End of Practice Test 1* ⭐

Great job finishing the test!

 *My Score*

I got ___________ out of 30 questions right.

📊 *Check Your Score Online!*

*Visit **ViewMath Academy** to enter your answers and see which topics you need to review. You can also explore lessons, take quizzes, track your scores, and save your progress!*

viewmath.com/score/3.1.NE.01

2

Practice Test 2

☑ 30 Questions

✏ Before You Start ✏

- ✓ **Read each question carefully** before choosing your answer.
- ✓ **Show your work** on scratch paper when you need to.
- ✓ **Skip hard questions** and come back to them later.
- ✓ **Check your answers** when you're done.
- ✓ **Take your time** — there's no rush!

⭐ You've Got This! ⭐

Write the expanded form of 9,205.

Maria rounded 438 to the nearest 10 and got 430. Is she correct?

A   *No, it should be 440*

B   *Yes, 438 rounds to 430*

C   *No, it should be 400*

D   *No, it should be 450*

3. *List all the even numbers between 21 and 30.*

How many hundreds are in 10,000?

Which is the correct expanded form of 720,056?

A $72,000 + 56$

B $700,000 + 20,000 + 50 + 6$

C $700,000 + 2,000 + 56$

D $700,000 + 20,000 + 56$

6. *A plane flew 1,892 miles on Saturday and 2,347 miles on Sunday. How far did it fly in total?*

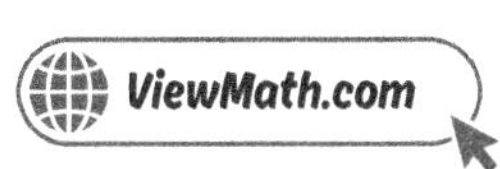

7. *What is* $5,000 - 2,346$?

  A) 2,544

  B) 2,654

  C) 2,754

  D) 3,654

8. *Estimate* $3,412 + 2,678$ *by rounding each number to the nearest thousand.*

  A) 5,000

  B) 6,000

  C) 7,000

  D) 6,090

9. *What is* $72 \div 9$?

  A) 7

  B) 8

  C) 9

  D) 63

10. *Write the missing fact in this fact family:* $7 \times 6 = 42,\ 6 \times 7 = 42,\ 42 \div 7 = 6,$ ___

  A) $42 \div 6 = 7$

  B) $42 \div 42 = 1$

  C) $7 + 6 = 13$

  D) $42 - 7 = 35$

11. *Which fraction means "two-thirds"?*

  A) $\frac{3}{2}$

  B) $\frac{2}{3}$

  C) $\frac{1}{3}$

  D) $\frac{2}{2}$

12. *A number line goes from 0 to 1 and is divided into 6 equal parts. Where is* $\frac{4}{6}$?

  A) 2 *jumps from* 0

  B) 3 *jumps from* 0

  C) 4 *jumps from* 0

  D) 5 *jumps from* 0

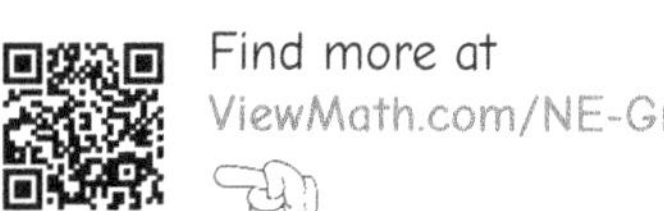
Find more at
ViewMath.com/NE-Grade3

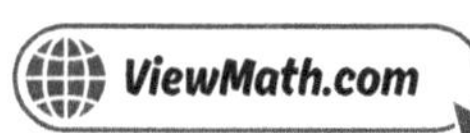

13. A ribbon is cut into 8 equal pieces. Sarah takes 1 piece. What fraction of the ribbon does she have?

A $\frac{8}{1}$ B $\frac{1}{8}$

C $\frac{7}{8}$ D $\frac{8}{8}$

14. Mia colors $\frac{3}{8}$ of a strip. How many more eighths does she need to color to reach $\frac{7}{8}$?

A 3   B 4

C 5   D 7

15. Which pair of fractions are equivalent?

A $\frac{1}{4}$ and $\frac{2}{8}$ B $\frac{1}{4}$ and $\frac{3}{8}$

C $\frac{2}{3}$ and $\frac{3}{4}$ D $\frac{1}{3}$ and $\frac{1}{4}$

16. Which of these is a way to write 5 as a fraction?

A $\frac{5}{1}$ B $\frac{10}{2}$

C $\frac{15}{3}$ D All of the above

17. Is $\frac{5}{8}$ more or less than $\frac{1}{2}$?

A Less than $\frac{1}{2}$ B Equal to $\frac{1}{2}$

C More than $\frac{1}{2}$ D Cannot tell

18. What is $\frac{1}{4} + \frac{1}{4} + \frac{1}{4}$?

19. What is $\frac{7}{8} - \frac{4}{8}$?

20. A marker reaches from 0 to the half-inch mark past 6 inches. How long is the marker?

21. A melon has a mass of 2 kg. A bunch of grapes has a mass of 500 g. How much heavier is the melon than the grapes?

   A) 498 g

   B) 1,000 g

   C) 1,500 g

   D) 2,500 g

22. Container A holds 20 liters. Container B holds 13 liters. How much more does Container A hold?

   A) 7 L

   B) 13 L

   C) 20 L

   D) 33 L

23. Lily has $6.00. She buys a sticker pack for $2.35 and a pen for $1.80. Does she have enough money left to buy a snack for $2.00?

   A) Yes, she has $2.85 left

   B) Yes, she has $2.00 left

   C) No, she only has $1.85 left

   D) No, she only has $1.65 left

24. A picture graph tracks books read in a month. Each symbol stands for 2 books. Amy has 5 symbols, Ben has 8 symbols, and Chloe has 3 symbols. How many books did they read altogether?

Your Answer

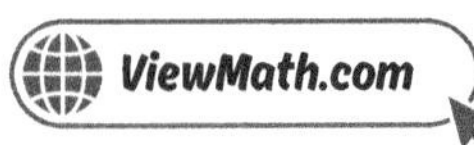

A line plot shows the lengths of 10 crayons. The data is shown below:

	2 in.	$2\frac{1}{2}$ in.	3 in.	$3\frac{1}{2}$ in.	4 in.
	2	3	4	1	0

How many crayons are shorter than 3 inches?

A) 2

B) 4

C) 5

D) 9

It is certain that the sun will set tonight. What does "certain" mean?

A) It might happen.

B) It will definitely happen.

C) It probably won't happen.

D) It can never happen.

27. How many sides does a hexagon have?

A) 5

B) 6

C) 7

D) 8

A square has sides of 9 inches. What is the area?

Which fraction is bigger: $\frac{1}{3}$ or $\frac{1}{6}$?

A) $\frac{1}{6}$ is bigger

B) $\frac{1}{3}$ is bigger

C) They are the same size.

D) You cannot tell.

Find more at
ViewMath.com/NE-Grade3

30. *Is a circle a polygon? Explain why or why not.*

Your Answer

⭐ *End of Practice Test 2* ⭐

Great job finishing the test!

📋 *My Score*

I got __________ out of 30 questions right.

📊 *Check Your Score Online!*

*Visit **ViewMath Academy** to enter your answers and see which topics you need to review. You can also explore lessons, take quizzes, track your scores, and save your progress!*

viewmath.com/score/3.1.NE.02

Practice Test 3

☑ 30 Questions

✏ Before You Start ✏

✓ **Read each question carefully** before choosing your answer.

✓ **Show your work** on scratch paper when you need to.

✓ **Skip hard questions** and come back to them later.

✓ **Check your answers** when you're done.

✓ **Take your time** — there's no rush!

★ You've Got This! ★

Each place value is how many times bigger than the place to its right?

A 2 times

B 5 times

C 10 times

D 100 times

2. When you round 549 to the nearest 10 and to the nearest 100, which gives the larger answer?

A Rounded to the nearest 10

B Rounded to the nearest 100

C They are the same

D Cannot tell without rounding

3. If you add two odd numbers, the result is always:

A Odd

B Even

C Greater than 10

D An odd number less than 20

4. What number is 1 more than 9,999?

A 9,000

B 9,990

C 10,000

D 10,001

5. What is the smallest 6-digit number?

6. What is $1{,}234 + 3{,}452$?

A 4,586

B 4,686

C 4,786

D 5,686

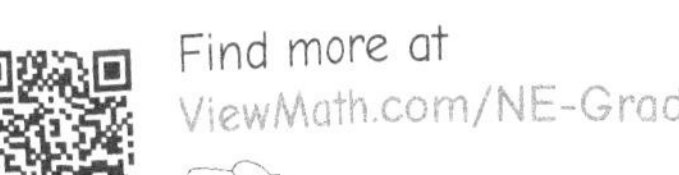
Find more at
ViewMath.com/NE-Grade3

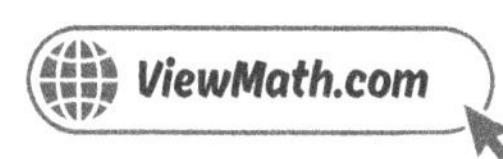

7. *What is $9{,}876 - 4{,}532$?*

   (A) 5,244

   (B) 5,334

   (C) 5,344

   (D) 5,444

8. *Estimate $724 - 389$ by rounding each number to the nearest hundred.*

   (A) 200

   (B) 300

   (C) 400

   (D) 335

9. *What is $35 \div 5$?*

   (A) 5

   (B) 6

   (C) 7

   (D) 8

10. *Which multiplication fact helps solve $56 \div 8$?*

    (A) $8 \times 6 = 48$

    (B) $8 \times 7 = 56$

    (C) $8 \times 8 = 64$

    (D) $8 \times 9 = 72$

11. *A shape is divided into 4 parts, but the parts are NOT equal. Can you write a fraction for the shaded part?*

    (A) Yes, it is $\frac{1}{4}$

    (B) Yes, any shaded part is a fraction

    (C) No, fractions require equal parts

    (D) No, you need 8 parts for a fraction

12. *Write the missing fractions:* $\frac{0}{4}, \frac{1}{4}, \underline{\quad}, \frac{3}{4}, \frac{4}{4}$

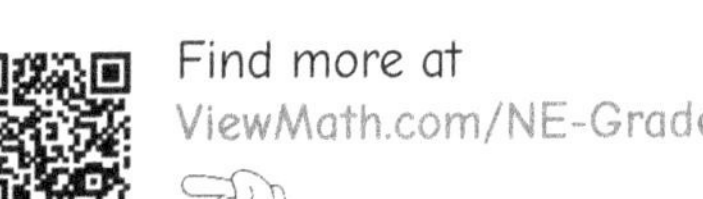

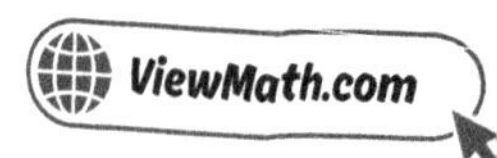

On a number line divided into 4 equal parts, where is $\frac{1}{4}$?

A At 0

B At the first mark after 0

C At the halfway point

D At 1

Alex says $\frac{3}{4}$ is 3 copies of $\frac{1}{3}$. Is Alex correct?

A Yes

B No, $\frac{3}{4}$ is 3 copies of $\frac{1}{4}$

C No, $\frac{3}{4}$ is 4 copies of $\frac{1}{3}$

D No, $\frac{3}{4}$ is 3 copies of $\frac{3}{3}$

15. Write two fractions that are equivalent to $\frac{1}{4}$.

On a number line divided into fourths, which fraction is at 3?

A $\frac{3}{4}$

B $\frac{4}{3}$

C $\frac{12}{4}$

D $\frac{3}{1}$

When two fractions have the same denominator, how do you compare them?

A The one with the bigger denominator is larger

B The one with the bigger numerator is larger

C They are always equal

D You cannot compare them

18. What is $\dfrac{2}{3} + \dfrac{1}{3}$?

A $\dfrac{3}{6}$

B $\dfrac{3}{3}$

C $\dfrac{2}{3}$

D $\dfrac{1}{3}$

Find more at
ViewMath.com/NE-Grade3

ViewMath.com

19. Liam says $\frac{8}{10} - \frac{3}{10} = \frac{5}{0}$. What did Liam do wrong?

   A. He added instead of subtracting.

   B. He subtracted the denominators too.

   C. He used the wrong numerators.

   D. He did nothing wrong.

20. A ribbon is $5\frac{1}{2}$ inches long. A piece of string is $4\frac{3}{4}$ inches long. Which is longer?

   A. The ribbon

   B. The string

   C. They are the same length

   D. Not enough information

21. Convert 3 kilograms to grams.

22. What unit do we use to measure liquid volume?

   A. Grams

   B. Kilograms

   C. Liters

   D. Inches

23. What is $\$3.45 + \$2.80$?

   A. $5.25

   B. $6.25

   C. $5.65

   D. $6.15

24. A picture graph shows toys collected. Each symbol stands for 5 toys.

Bears: ★★★
Cars: ★★★★★
Dolls: ★★

How many toys were collected in all?

A 10

B 25

C 50

D 75

25. A line plot shows the lengths of nails in inches. The data points are: 1, 1, $1\frac{1}{2}$, 2, 2, 2, $2\frac{1}{2}$, $2\frac{1}{2}$. How many X marks should be placed above 2 inches?

26. You flip two coins. Which outcome is possible?

A Both heads

B Both tails

C One head and one tail

D All of the above

27. Explain why a square is a special kind of rectangle AND a special kind of rhombus.

28. A rectangle is 10 feet long and 3 feet wide. What is its area?

A 13 sq ft

B 26 sq ft

C 30 sq ft

D 33 sq ft

Find more at
ViewMath.com/NE-Grade3

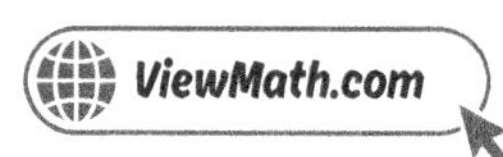

29. A square is partitioned into 4 equal parts. All 4 parts are shaded. What fraction is shaded?

(A) $\frac{1}{4}$ (B) $\frac{3}{4}$

(C) $\frac{4}{4}$ (D) $\frac{4}{1}$

30. A polygon has 3 more sides than a pentagon. What polygon is it?

(A) Hexagon   (B) Heptagon

(C) Octagon   (D) Nonagon

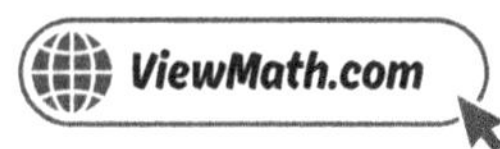

⭐ End of Practice Test 3 ⭐

Great job finishing the test!

 My Score

I got ____________ out of 30 questions right.

📊 Check Your Score Online!

*Visit **ViewMath Academy** to enter your answers and see which topics you need to review. You can also explore lessons, take quizzes, track your scores, and save your progress!*

viewmath.com/score/3.1.NE.03

Answer Key & Explanations

Check Your Answers!

First try each test on your own, then look here to check.

Read the explanations to learn from any mistakes

✅ Practice Test 1 — Answer Key

1. $5{,}000 + 600$ 2. C 3. B 4. B 5. D 6. $10{,}245$ 7. $3{,}636$ 8. B

9. 72 10. C 11. C 12. B 13. $\frac{1}{5}$ 14. C 15. 1 16. D 17. B 18. $\frac{4}{4}$ or 1

19. B 20. B 21. C 22. $3\,L$ 23. B 24. D 25. A 26. D 27. C 28. D

29. A 30. C

💡 Time to Learn! 💡

Go through the explanations below, especially for the questions you missed.

Understanding why each answer is correct makes you a stronger math thinker!

👍 Tip: Circle any questions you got wrong, then read their explanation carefully.

📖 Practice Test 1 — Detailed Explanations

1. $5{,}600 = 5{,}000 + 600 + 0 + 0$. The tens and ones are both 0.

Find more at
ViewMath.com/NE-Grade3

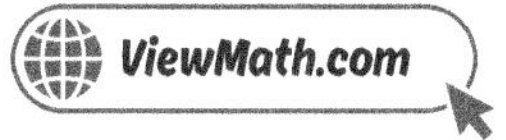

462 has tens digit $6 \geq 5$, so it rounds up to 500. 428 and 449 round to 400. 551 rounds to 600.

20 (ends in 0), 46 (ends in 6), 88 (ends in 8) are all even. The other groups each contain at least one odd number.

$8,000 + 2,000 = 10,000.$

In 283,041, the digit 8 is in the ten-thousands place, so its value is 80,000.

Ones: $9 + 6 = 15$, carry 1. Tens: $8 + 5 + 1 = 14$, carry 1. Hundreds: $7 + 4 + 1 = 12$, carry 1. Thousands: $6 + 3 + 1 = 10$. The sum is 10,245 — a 5-digit number! Two 4-digit numbers can add up to more than 9,999.

$7,204 - 3,568 = 3,636.$ Ones: $4 < 8$, borrow through zero: $14 - 8 = 6$. Tens: $9 - 6 = 3$. Hundreds: $1 - 5$, borrow: $11 - 5 = 6$. Thousands: $6 - 3 = 3$.

$347 \approx 300$ and $582 \approx 600$. The estimate is $300 + 600 = 900$, which is less than the exact sum of 929.

If the number $\div 8 = 9$, then the number $= 9 \times 8 = 72$.

$4 \times 5 = 20$, not 25. So $4, 5, 25$ do not form a fact family.

The numerator (top number) tells how many parts you have.

$\frac{2}{4}$ is at the second of 4 marks, which is the middle of the line from 0 to 1.

5 equal shares means each person gets $\frac{1}{5}$ of the sandwich.

$\frac{1}{3} + \frac{1}{3} + \frac{1}{3} = \frac{3}{3} = 1$ whole. It takes 3 copies.

Find more at
ViewMath.com/NE-Grade3

Divide top and bottom by 2: $\frac{2 \div 2}{4 \div 2} = \frac{1}{2}$.

$4 \times 3 = 12$ thirds. So $4 = \frac{12}{3}$.

Same denominator. $5 > 3$, so $\frac{5}{8} > \frac{3}{8}$. Mia ate more.

$\frac{3+1}{4} = \frac{4}{4} = 1$ whole.

To subtract fractions with like denominators, subtract the numerators and keep the denominator the same.

$6\frac{3}{4} - 6\frac{1}{4} = \frac{3}{4} - \frac{1}{4} = \frac{2}{4} = \frac{1}{2}$ inch.

$3 + 5 = 8$ kg.

$4 - 1 = 3$ L.

Line up the decimals and add: $\$2.50 + \$1.25 = \$3.75$.

Each star $= 5$ books. Jake has 3 stars, so $3 \times 5 = 15$ books.

Each X mark represents one pencil. 3 X marks above $1\frac{1}{2}$ means 3 pencils measured $1\frac{1}{2}$ inches.

There are no blue marbles in the bag, so picking a blue marble is impossible.

A square always has 4 equal sides and 4 equal angles ($90°$ each), so it is always regular. A rectangle has equal angles but not always equal sides.

A square is a rectangle with all sides equal. Area $= 7 \times 7 = 49$ sq in.

Find more at
ViewMath.com/NE-Grade3

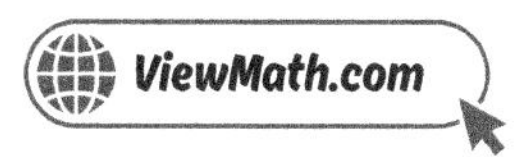

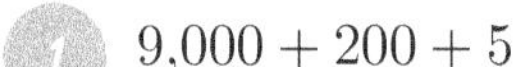 When a whole is split into 2 equal parts, each part is $\frac{1}{2}$ (one-half).

A STOP sign is a regular octagon with 8 equal sides.

✅ Practice Test 2 — Answer Key

1 $9,000 + 200 + 5$ **2** A **3** $22, 24, 26, 28, 30$ **4** 100 **5** B **6** $4,239$ **7** B

8 B **9** B **10** A **11** B **12** C **13** B **14** B **15** A **16** D **17** C

18 $\frac{3}{4}$ **19** $\frac{3}{8}$ **20** $6\frac{1}{2}$ inches **21** C **22** A **23** C **24** 32 **25** C **26** B

27 B **28** 81 sq in **29** B **30** No, because a circle has curved sides, not straight sides

💡 Time to Learn! 💡

Go through the explanations below, **especially for the questions you missed.**

Understanding why each answer is correct makes you a stronger math thinker!

👍 *Tip:* Circle any questions you got wrong, then read their explanation carefully.

📖 Practice Test 2 — Detailed Explanations

1 $9,205 = 9,000 + 200 + 0 + 5$. The tens digit is 0, so there is no tens term.

2 The ones digit is 8. Since $8 \geq 5$, we round up. 438 rounds to 440, not 430.

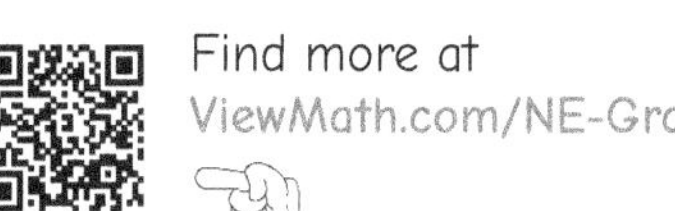 Find more at
ViewMath.com/NE-Grade3

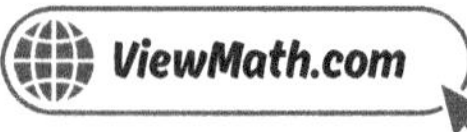 ViewMath.com

The even numbers between 21 and 30 are 22, 24, 26, 28, 30. They all end in an even digit.

$10,000 \div 100 = 100$. There are 100 hundreds in 10,000.

$720,056 = 700,000 + 20,000 + 0 + 0 + 50 + 6$. The thousands and hundreds places are both zero.

$1,892 + 2,347 = 4,239$. Ones: $2 + 7 = 9$. Tens: $9 + 4 = 13$, carry 1. Hundreds: $8 + 3 + 1 = 12$, carry 1. Thousands: $1 + 2 + 1 = 4$.

Borrow through all the zeros: $10 - 6 = 4$, $9 - 4 = 5$, $9 - 3 = 6$, $4 - 2 = 2$. The answer is 2,654.

$3,412 \approx 3,000$ and $2,678 \approx 3,000$. So $3,000 + 3,000 = 6,000$.

Think: $9 \times ? = 72$. Since $9 \times 8 = 72$, the answer is 8.

The missing fact is $42 \div 6 = 7$. A fact family with different factors has 4 facts.

"Two-thirds" means 2 parts out of 3 equal parts $= \frac{2}{3}$.

$\frac{4}{6}$ means 4 jumps from 0 on a number line split into 6 equal parts.

She takes 1 piece out of 8 equal pieces $= \frac{1}{8}$.

$\frac{7}{8} - \frac{3}{8} = \frac{4}{8}$. She needs 4 more eighths.

$\frac{1}{4} = \frac{1 \times 2}{4 \times 2} = \frac{2}{8}$.

$\frac{5}{1} = 5$, $\frac{10}{2} = 5$, and $\frac{15}{3} = 5$. They all equal 5.

Find more at
ViewMath.com/NE-Grade3

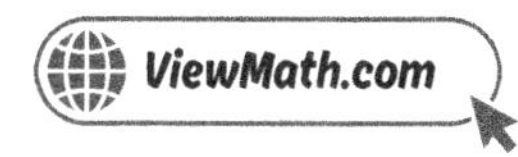

$\frac{1}{2} = \frac{4}{8}$. Since $\frac{5}{8} > \frac{4}{8}$, it is more than $\frac{1}{2}$.

$\frac{1+1+1}{4} = \frac{3}{4}$.

$\frac{7-4}{8} = \frac{3}{8}$.

The marker ends at the half-inch mark past 6, which is $6\frac{1}{2}$ inches.

$2\ kg = 2{,}000\ g$. Then $2{,}000 - 500 = 1{,}500\ g$.

$20 - 13 = 7\ L$.

Total spent: $\$2.35 + \$1.80 = \$4.15$. Money left: $\$6.00 - \$4.15 = \$1.85$. Since $\$1.85 < \$2.00$, she does not have enough.

Amy: $5 \times 2 = 10$. Ben: $8 \times 2 = 16$. Chloe: $3 \times 2 = 6$. Total: $10 + 16 + 6 = 32$ books.

Crayons shorter than 3 inches: 2 in. (2) + $2\frac{1}{2}$ in. (3) = 5 crayons.

Certain means it will definitely happen — there is no doubt.

A hexagon has 6 sides and 6 vertices. "Hex" means 6.

Area $= 9 \times 9 = 81$ sq in.

When the whole is the same size, fewer parts means bigger pieces. $\frac{1}{3}$ means 3 parts, $\frac{1}{6}$ means 6 parts. $\frac{1}{3} > \frac{1}{6}$.

A polygon must have straight sides. A circle is curved, so it is not a polygon.

Find more at
ViewMath.com/NE-Grade3

✅ Practice Test 3 — Answer Key

1. C	2. A	3. B	4. C	5. 100,000	6. B	7. C	8. B	9. C
10. B	11. C	12. $\frac{2}{4}$	13. B	14. B	15. $\frac{2}{8}$ and $\frac{3}{12}$	16. C	17. B	18. B
19. B	20. A	21. 3,000 g	22. C	23. B	24. C	25. 3	26. D	

A square has 4 right angles (like a rectangle) and 4 equal sides (like a rhombus), so it belongs to both families.

27. C   28. C   29. C

💡 Time to Learn! 💡

Go through the explanations below, *especially for the questions you missed.*

Understanding why each answer is correct makes you a stronger math thinker!

👍 *Tip:* Circle any questions you got wrong, then read their explanation carefully.

📖 Practice Test 3 — Detailed Explanations

Each place is 10 times bigger than the place to its right: ones → tens → hundreds → thousands.

549 rounded to the nearest 10: ones digit is 9 ≥ 5, so 550. Rounded to the nearest 100: tens digit is 4 < 5, so 500. 550 > 500.

Odd + Odd = Even. For example, 3 + 5 = 8 (even) and 7 + 9 = 16 (even).

9,999 + 1 = 10,000. This is the first 5-digit number.

Find more at
ViewMath.com/NE-Grade3

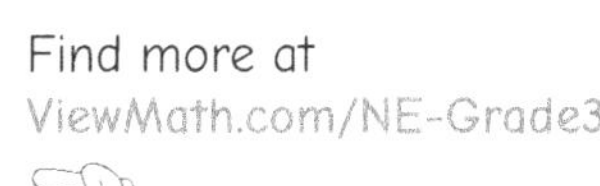

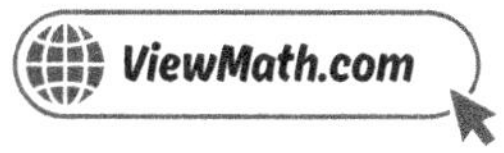

The smallest 6-digit number is 100,000. Any number below it (like 99,999) has only 5 digits.

Ones: $4 + 2 = 6$. Tens: $3 + 5 = 8$. Hundreds: $2 + 4 = 6$. Thousands: $1 + 3 = 4$. The sum is 4,686.

No borrowing is needed. Ones: $6 - 2 = 4$. Tens: $7 - 3 = 4$. Hundreds: $8 - 5 = 3$. Thousands: $9 - 4 = 5$. The difference is 5,344.

$724 \approx 700$ and $389 \approx 400$. So $700 - 400 = 300$.

Skip count by 5s: $5, 10, 15, 20, 25, 30, 35$. That's 7 jumps. So $35 \div 5 = 7$.

$8 \times 7 = 56$, so $56 \div 8 = 7$.

Fractions only work when the whole is divided into **equal** parts.

Counting fourths in order: $\frac{0}{4}, \frac{1}{4}, \frac{2}{4}, \frac{3}{4}, \frac{4}{4}$.

The unit fraction $\frac{1}{4}$ is always at the first mark after 0 on a number line divided into 4 parts.

$\frac{3}{4}$ means 3 copies of $\frac{1}{4}$, not $\frac{1}{3}$. The denominator tells the size of each piece.

$\frac{1 \times 2}{4 \times 2} = \frac{2}{8}$ and $\frac{1 \times 3}{4 \times 3} = \frac{3}{12}$.

$3 = 3 \times \frac{4}{4} = \frac{12}{4}$. On a fourths number line, 3 is at $\frac{12}{4}$.

Same denominator means same-size pieces. More pieces (bigger numerator) = bigger fraction.

$\frac{2 + 1}{3} = \frac{3}{3}$, which equals 1 whole.

 Liam subtracted both the numerators and the denominators. The denominator should stay the same: $\frac{8-3}{10} = \frac{5}{10}$.

$5\frac{1}{2} = 5\frac{2}{1}$, which is greater than $4\frac{3}{1}$. The ribbon is longer.

$3 \times 1{,}000 = 3{,}000$ g.

We measure liquid volume in liters (L).

Cents: $45 + 80 = 125$ cents $= 1$ dollar and 25 cents. Dollars: $3 + 2 + 1 = 6$. Answer: $6.25.

Bears: $3 \times 5 = 15$. Cars: $5 \times 5 = 25$. Dolls: $2 \times 5 = 10$. Total: $15 + 25 + 10 = 50$ toys.

Count how many times 2 appears in the data: 2, 2, 2. That's 3 times.

When flipping two coins, you can get heads-heads, tails–tails, or one of each. All outcomes are possible.

A rectangle needs 4 right angles — a square has those. A rhombus needs 4 equal sides — a square has those too. So a square is both.

Area $= 10 \times 3 = 30$ sq ft.

All 4 out of 4 parts are shaded, so $\frac{4}{4}$ is shaded. $\frac{4}{4} = 1$ whole.

A pentagon has 5 sides. $5 + 3 = 8$ sides. An octagon has 8 sides.

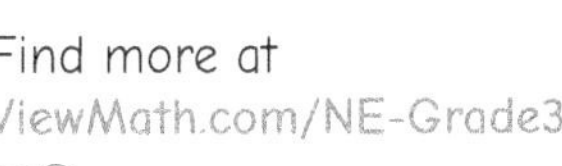

Find more at
ViewMath.com/NE-Grade3

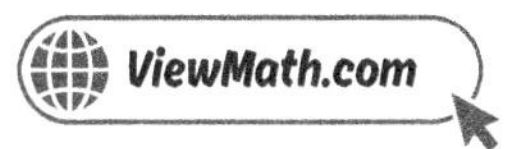

Great job checking your work!

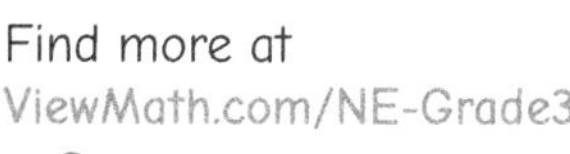
Find more at
ViewMath.com/NE-Grade3